How to
Build a Bow Top

BY

Walter Lloyd

TRAVELLER EDUCATION SERVICE
SOUTH GLAMORGAN
GREENWAY SCHOOL
LLANSTEPHAN ROAD, RUMNEY   TEL. 790211

Published by Woodmanship Ltd
PO Box 12, Carnforth, Lancashire, LA5 9NN
ISBN 09535238 0 2
© 1999 Walter Lloyd
Printed by Kent Valley Colour Printers, Kendal, Cumbria

How to build a Bow Top

The Living Wagon or Gipsy Caravan developed during the 19th century, as various travellers, including showmen and Gipsies found that they could build a horse drawn living accommodation to take advantage of the improved road surfaces and the bigger horses. At first it consisted of just a canvas tilt on a two- or four-wheeled cart. One of the early examples was the four-wheeled 'Pot Cart' that was given an accommodation with a light framework that could be lifted off and lived in while the cart was used during the day for hawking and so on. At first a simple bender with the rods fastened to wooden boards instead of driven into the ground – this developed into the bow-top wagon of the present day.

As this idea caught on, wheelwrights and coach builders got in on the act and started to build more advanced types – the Showman or Burton wagon, the Bill Wright or Ledge wagon, and the Reading wagon and rare Brush wagon. Some purists would say that the term Bow Top is only used for large wagons that have their rear wheels coming up outside the floor but under the seats. There are so many different styles of wagon that it is difficult to classify them accurately.

Today, although many wagons are used to attend horse-fairs throughout the summer, they are mostly home-built by people who are either not from traveller families, or if they are, they have given up travelling all the year round. A recent development is that people who are tired of inner city life with its shortage of work and high levels of stress, decide to take to the road and find that travelling with horses is in many ways more satisfactory than using motors. This is not to say that life on the road with horses is easy – it isn't, but it has advantages. Many families and individuals travelling with horses can be seen at Horse Fairs such as Appleby, Stow on the Wold, Priddy and others. At Appleby in recent years there have probably been over 100 wagons, but not all of these camped on Fair Hill, many in the lanes and on common land up to 20 miles away, or in rented fields near the Fair. The best place to see them is probably the road from Kirkby Lonsdale to Sedbergh and on towards Kirkby Stephen at the time of Appleby Fair, which finishes on the second Wednesday in June.

The intention of this booklet is to describe how to build a simple Bow-Top or Open-Lot Living Wagon. A Bow-Top is light to pull, simple to build, economical in materials, but also warm and comfortable, even in winter. It can be pulled by almost any size horse (or horses), but it is always important to keep the weight to a minimum for the sake of the horse, whatever size.

Before you start to build, it is important to have experience of driving horses. If you don't already have this experience, get a pony and flat cart and travel with them for a bit to learn about the road, and only then think about a wagon. If you have never owned a horse, let your first one be the cheapest – the first is unlikely to be the right one anyway, and a cheap one means that you will lose less money when you come to sell it or swop it for another.

The Dray

So – you have been travelling with the pony and cart all summer and you plan to build a bow top during the winter so that you can travel with it next summer. The first thing is to get a good dray/lorry/rully/ trolley – all names for a four wheel spring cart with a flat floor. The standard size is 9ft long by 5 feet wide, with the floor about 3 ft off the ground. Bigger than this will be heavy to pull, and unless you are only going to travel in country where there are no hills, forget it. Smaller is all right, 8ft long by 4ft is fine. Smaller than that is getting a bit small to live in. Wooden wheels are far better than motor wheels; iron tyres may make a lot of noise but they are far easier to pull on tarmac than motor tyres, or even than solid rubber tyres. If you can't get, or can't afford wooden wheels, then go for the three stud 19 inch so-called artillery wheels (real artillery wheels are wooden and are designed to be repaired at the road side by inserting new spokes or fellies, but they are very rarely seen nowadays). Failing artillery wheels, 'easy clean' are a possibility.

The dray will have a forecarriage consisting of the turntable with a pair of large iron rings, one on the turntable and one on the body of the dray, turning on each other to give a full lock, springs and wheels and axle and wooden axle bed (some don't have the wooden bed).

The forecarriage is fixed to the floor of the dray with a big iron King Pin that allows it to turn with the two rings in contact.

The floor consists of (usually) four 'summers', wooden spars that run front to back. The rear springs are fastened either to the outer summers or in the case of the Evesham, West Country type, to two separate spars that are separated from the summers by two or three cross pieces. The springs in turn are fastened to the rear axle or axle bed.The floor boards run from side to side across the summers. On top of the floor boards, and running all round the outer edge are the choc rails. There may be a tail board at the back, with quarter rails on top of the side choc rails for a couple of feet at the back. There may be a foot board in front of the front choc rail, giving extra length. Drays are not always sold with shafts, but you will need them if you are going to drive a horse. If not sold with the dray, they will be expensive to buy separately, probably at least another hundred pounds. The traction should be by trace hooks set at the rear of the shafts, and ideally there should be iron work all the way from the trace hooks to the shaft rings that fit on the shaft pin, linking them to the splinter bar on the fore carriage. Some shafts are fitted for heavy horse harness, but this is almost never used with a light bow top. Whatever dray you get, new or second hand, make sure it has good brakes, with a brake wheel that you can reach easily whether walking beside it or sitting up at the front of the seat. Your life, and that of the horse, might depend on the brakes! Very good quality breeching straps on the harness are important for the same reason.

Second hand drays are not easy to find. It is sometimes cheaper to buy a second hand bow top than a dray. At 1998 prices a good new dray can cost £1250 to £1850. If you go to a wheelwright for a new one there is probably a waiting list, although there are often good new drays at horse fairs such as Huddersfield (Easter Monday), Appleby (early June) and Lee Gap (14 August and 17 September). I like to have a 'Chesterfield' dray built to my specifications by Sonny Thorpe at Temple Normanton, tel 01246 864861 He usually travels up to Appleby Fair 'by road' as they say, meaning driving horses, and he knows from experience what is needed for that sort of work. His drays are very light.

The Top

You now have your dray, so you have to design a top to go on it. Virtually all the wagons on the road are different – no two the same, but they are nearly all within a clear tradition. I sometimes see wagons built by someone who thinks that a lot of modern innovations will be more efficient – things like disc brakes, tubular steel and so on. They invariably fail and are never seen again. Wagons built by newcomers to the business are very often far too heavy, so that the horses have serious trouble on the hills and end up with sore shoulders.

The elements of a Bow Top are:-

A. The **sides, back and front**, made of planks about a foot wide. These are either fastened on top of the choc rails, or sit on the floor just inside the choc rails – especially if the top is designed as an accommodation to be taken off so that the dray can be used separately.

B. Screwed on top of the sides and back and front are the **seats**, running full length of the sides.

C. Fastened to the front and back boards are the **king posts**, two (or sometimes four) at the front, two at the back, with the **crown boards** fastened to them at the top. The position and shape of the crown boards are crucial for a good shaped Top. The back is filled in between the rear crown board and the back board with planking, often the type called penny boarding, narrow planking with tongue and groove, to turn the weather. This is sometimes called the **end deck**. The **window** is in the end deck below the crown board.

E. At the front are the two **front formers**, narrow curved planks, that follow the same curve as the end deck to give the roof its shape.

F. The **runners** are the full length of the roof, let into tight fitting slots in the front and rear crown boards, usually five or seven or nine of them, with extra ones sometimes extending down the sides to give a more even shape. They should be longer than the seats to give an overhang effect.

G. The sides are then **boarded** up for 18 inches to two feet above the seats, extended beyond the end deck and front formers, as are the runners, to give the overhang.

H. Now come the **bows**, usually ten or twelve of them according to the length of the top. They are bent over the front and rear crown boards and the runners, with one in front of the front crown board, and one behind the rear crown board.

I. This is the time to build the **bed**, before the sheet is on, so that you can maneover full width planks into place. The bed helps to give the top its shape, and adds to the strength.

J. Now at last you can see what the finished wagon is going to look like, ready to be covered with a **lining**, **insulation**, and a canvas **sheet**, and **front sheets**.

K. The **weatherboards** that cover the edges of the canvas sheet, one each side, one at the back, one at the front.

L. The **pan box** fastened underneath the dray between the back wheels, for storing things like the soot blackened kettle, harness oil, paraffin for hurricane lamps and so on.

M. The **Cratch** or **heck**, a rack hinged at the bottom where it is attached to the back of the wagon and used for storing harness, hay and other bulky things.

N. This in turn is covered by a **cratch sheet** of canvas, fastened onto the end deck just below the window, usually with a narrow board, and tied down over the loaded cratch.

P. For the **paint**. Really a separate subject in itself, often done by specialists.

Named parts of a Bow Top wagon

Bows
runners
front crown board
rear crown board
window
king post
end deck
front former
side planking
bearer for bed
bed frame
back board
seat
side
front board
choc rail
brake wheel
fore carriage
splinter bar
axle bed

Photo 1

This is an accommodation top that I built in 1993, on an eight foot by four foot Chesterfield dray built by Sonny Thorpe in 1986. Overall length at the seats eight foot eight, width at the seats, five foot four. Length inside from end deck to front former a bare six feet. The bed extends from two feet nine to four feet five. Weight probably about seven hundredweight. I weighed an earlier, full size wagon, empty, and it was exactly 10cwt.

Design

I like to draw the shape of the front or back of the wagon full scale on a wall, to get the shape just right. I have spent a full week drawing and modifying this shape before I was satisfied that it was right, and could start cutting wood.

I mark on the wall the position of the floor, full width and the right height above the ground. Mark in the choc rails, and draw the sides, either sitting on the choc rails, or on the floor inside the choc rails. I usually use 12 inch wide planks, as it is the largest that it easy to get, but I have used 15 inch when I could get them, to give higher seats. I always use **parana pine** for the sides, back and front boards, seats, king posts and crown boards, as it is light, strong, and very straight grained with practically no knots. One inch is too thick, adding unnecessary weight, and half inch is not quite strong enough, five eigths or three quarter inch would be about right (planed to that thickness, finished – timber merchant's sizes can be confusing as sometimes they quote the size before planeing.) I have been getting parana pine from James Ashworth (Waterfoot) Successors Ltd Tel. Rossendale 01706 214484. At the time of writing, (April 1998) thay have it in stock in 12 inch widths, but only in 1 inch thickness, "a lot of expensive sawdust", as they said, to bring it down to half inch. Price about £2.16 per foot plus VAT.

The seats are now marked on top of the sides, twelve inch again,or whatever you use for the sides, balanced so that the centre line of the seat sits on the side.

Now decide how high the roof is to be from floor level, and mark in the crown board at that height, and draw in the king posts to support it. Mark in the position of the bed, which will be across the back of the wagon. Unless you are very short, the bed will need to be over six foot across at top-of-the-mattress level. If your head and feet are going to be at the widest part of the wagon, the wooden bed frame will be the thickness of the mattress lower Now you draw a curve from the centre of the top of the crown, round and down past the bed, and finishing at the outer edge of the seat. As I said earlier, it once took me a week to get that curve right, with both sides identical, and

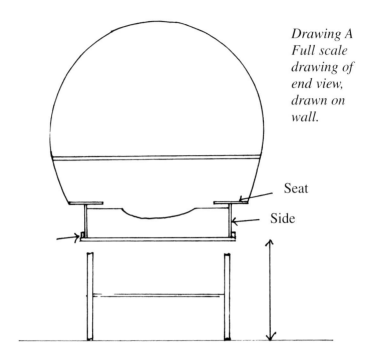

*Drawing A
Full scale
drawing of
end view,
drawn on
wall.*

Seat

Side

finally I did it. When you are really sure that the full size drawing is right, you can start actually building the top.

You may find that the floor of the dray is not exactly rectangular – one side can be half an inch to an inch longer than the other, and the same with front and back boards, so each piece of wood has to be individually measured and cut. Never forget the old joiners saying – 'measure twice and cut once.'

The sides should be up to 8 inches to a foot longer than the length of the dray floor, with the ends cut to a nice curve or double curve sweeping up to the seats. If the sides are to sit directly on the floor, accommodation fashion, this curve will allow the side to fit on the floor and project forwards and backwards, and also prevent the whole top from sliding as you go up or down hill.

If you are leaving the dray's own tailboard and quarter rails in place, some clever cutting has to be done for the side and seat to project beyond the tailboard.

The front and back boards sit tight against the sides, either on the floor or on the choc rails. If this is to be a Whitworth Waggon, the front board is set back from the front choc board, far enough to fit a gas bottle or water jack on the floor, and enough to give you a driving seat outside but sheltered from some of the weather. If this is an accommodation type top, with the sides sitting on the floor, it is possible to set the back board forward from the tail board by several inches to give storage for ropes, spare harness and so on. Of course, this reduces the space inside the wagon, but makes life much easier when on the move. I have just had the tailboard of my wagon modified, hinged at the bottom so that it opens backwards and acts as a heck. The front porch, in particular, is very handy for hanging wet clothes to dry, and for putting tea things and so on where you can reach them from outside, but safe from dogs, (Traditionally, the old Travelling People were very particular never to allow a dog to lick a plate or cup, they would even smash a crown derby plate, worth perhaps hundreds of pounds, if a dog licked it. They always used, and still use, two bowls, one for washing themselves and their clothes, a separate one, nowadays often of stainless steel, for washing the cups and plates and cutlery. This cleanliness tradition is very important to them. Equally important are:- never throw any rubbish onto a cooking fire, and don't walk between the fire and someone sitting beside it). If the sides are to sit on the floor inside the choc rail, it is possible to lay a sheet of lino first, so that the sides and front and back boards sit tight on the lino and help make all waterproof. If the sides are on top

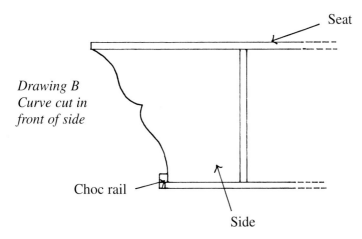

Seat

Drawing B
Curve cut in
front of side

Choc rail

Side

of the choc rail, but the front board is set back, Whitworth Wagon fashion, the front board will have to be the depth of the choc rail deeper to fit tight between both floor and seat.

The front and back boards should be a tight fit between the sides. I fasten them with brass counter sunk screws, usually one and a quarter or one and a half inch No. 8, predrilled and countersunk. The front board usually has a curve cut out to make it easier to step over.

The seats go on next, with the centre line of the seat sitting on the side, and screwed down, same method and screws, onto the sides and front and back boards. I curve the ends of the seats, with the longest side on the outside.

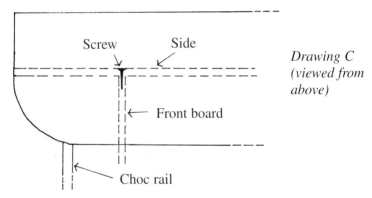

Drawing C (viewed from above)

As I always try to reduce weight by using the thinnest possible planking, I take the curves that I have cut from the front and back of the sides (cut with a jig saw), turn them upside down and using them as brackets, screw them to the sides to support the seats at the point of most stress, the driving position outside, and the usual sitting position in front of the bed. (DRAWING D). The sides of the wagon need to be fastened down to the dray. I usually fasten an accomodation top by just bolting through sides to choc rails, using one or two pairs of quarter inch coach bolts. If the sides are on top of the choc rails, they can be screwed to small metal plates joining the two.

Now comes the end deck. I use quarter inch or 6mm coach bolts ('cup square') to fasten the one and three quarter inch square (parana) king posts to the crown board, with the washers and nuts on the inside where I can see them if they work loose. I screw on the uncut penny boarding (I have usually used tongue and groove cedar, planed to just under half an inch) while the assembly is on the ground, with the tongue and groove in tight, bolt the whole thing onto the back

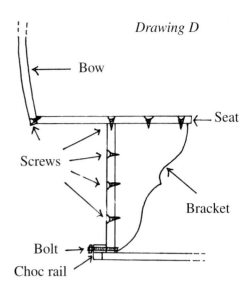

Drawing D

Bow

Seat

Screws

Bracket

Bolt

Choc rail

Photo 2. Back View. Sides, seats and end deck in position, rear crown not yet fitted. Note cratch or heck. The side board extends forwards of the floor with a nice indented curve. This is an 'accommodation' type wagon, with the side sitting on the floor, inside the choc rails and quarter rails.

11

board, and then mark in the curve and saw the ends of the penny boarding to the right shape, allowing for the side planks to be recessed to give a smooth curve on the outside. A bow will fit over this curve later, so it is important to get a smooth curve that will leave no gaps for wind and rain to get in. You may notice that in the photograph (no 2) I have fastened on the pennyboarding before the crown board.

Photo 3. Looking at the end deck from inside. Note poor fit with gaps on the seat, and recess for planking to fit on top of the seat.

The front is then assembled, but with no boarding, it can be put together on the wagon, King posts and crown board bolted with quarter inch coach bolts as before, but this time using wing nuts, they tend to work loose with the vibration of travelling, and you can reach and tighten them as you drive! The front formers are a bit tricky to fasten, I usually screw them to a square section that is in turn screwed to the seat, at the bottom, and at the top I butt them up to the crown board and carefully use long screws through the edge. I use the end deck as a pattern when cutting the curve, trying to get front and back identical.

Cutting the slots for the runners is tricky, too, to get a really tight fit. I like inch and a half by three quarters parana pine. They should be spaced fairly evenly across the crown boards, but if you are using extra ones, these will have to fit into slots in the end deck and front formers. Leave the runners overlong, to be cut to length later.

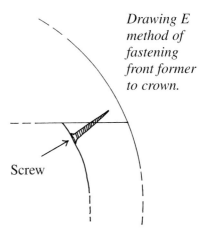

Drawing E method of fastening front former to crown.

Screw

The side planking can be very light, often tongue and grooved cedar as for the end deck, left overlong, to be cut to length later, after the bows are in place, screwed carefully with shorter, maybe thinner brass screws to the end deck and front formers where they touch. Many people like to have additional side planking at each

Photo 4.

Front and rear crown boards fitted. The runners are stacked against the roof of the shed, ready for fitting into the slots cut in the crown boards. Note the planking of the end deck is recessed to take the side boarding. You can see the space for the porch between the front board and the front choc rail.

13

*Photo 5
Runners.*

*Photo 6
Front formers*

end of the bed. adding another foot or two to the height of the planking here.

The bows are the most difficult part of the whole job. I use parana pine again, as it is so free of knots and straight grained. I am currently using inch and a half by a quarter inch or just over. The length will probably be fourteen feet or more, but it depends on the curve you have designed. I soak them in a beck or pond for about a week, to

Photo 7
Side planks

Photo 8
Bows

make them more flexible. I usually fasten the first one over the end deck, screwing it down to the top runner with a brass screw. Then I tie a piece of baler band or string to each end and gently pull them down, a little at a time, one end after the other and tie the other end of the band to the underworks. Now I pour boiling water from a kettle on the bow, pulling down one side at a time, until it is touching the next runner, when I screw it down, and then the other side to the runner on that side. More boiling water, add more gentle pressure until the bow is screwed down to all the runners, and the lower end screwed into the edge of the seat. it is not easy getting just the right curve, to match the designed curve on the end deck. Sometimes a bow kinks a bit, with a too sharp angle, spoiling the shape, but more often it breaks. You always need some spares in reserve, it takes a very good man to bend a set of bows without breaking any. Carry on until all the bows are in place, remembering that the front and back bows are going to be tilted, from the projecting top runner down to the front of the seats. Now you can tidy up all the ends, the ends of runners and of side planking can be cut flush with the front and back bows, and the ends of the bows can be cut off flush with the seat. It is useful to fasten the cut off bow ends along the edge of the seat, filling in between the bottom ends of the bows so that when the canvas sheet is fastened down, it has a flat surface to be fastened to.

Photo 9 Rear view

Parana the same size as the king posts is ideal for the bed frame. A sliding bed, that stows away during the day leaves more room for moving around in. Inside a wagon is pretty cramped anyway, so every little helps. There is a little extra weight, but I think it is worth it. First of all bolt the bearers to the side planking, using one and three quarter inch to two inch square, and about four foot eight long, placed so that the mattress top will be just below the widest part of the bows (in the wagon I am using now, that puts the top of the bed boards at thirty inches above the floor). Remember in the design stage that a lower bed will give you room to sit up in it without ducking your head, but a higher bed will give more storage space underneath. A typical width for a caravan bed will be four foot six, with the mattress in two halves lengthwise so that one half, slightly narrower, will fold over onto the other half during the day, bedding and all. Two cross bars are bolted to the bearers, one right at the back, close up to the rear king posts, the other about thirty inches from the king posts: these should be one and three quarters to two inches square, and relatively free from knots, which is why I again favour parana. Two similar cross bars are now placed loose on the bearers, one close up to the rear cross bar, the other outside and in front of the front cross bar. The bed boards, which can be of soft wood,three to four inches wide and half an inch thick, are screwed to the cross bars, first one between the bolted down bars, then one to the loose bars, with enough space between them to allow them to slide past each other, and so on to the other end. If I have written this in a way that you can understand, it should be possible to pull the front loose cross bar out to the full extent of the sliders, bringing alternate bed boards out attached to the rear loose bar, so nearly doubling the width of the bed. A two piece mattress can be folded back and the bed pushed in during the day, with the bedding on top.It is usual to screw a front board upright along the front loose cross bar, and the top of this is often chamfered or carved. This front board keeps the mattress in place when folded out

The main woodwork is now completed, and the top can be covered, first with a lining, usually a thin patterned material, and then some sort of insulation. Some people use carpet to provide both pattern and insulation, but carpet is very heavy. I use the wadding that is sold for insulating and padding anoraks. I staple both the lining and wadding into place, first pulling them down tight. Then the canvas sheet. Green

Photo 10 Inside front

(Willesden) canvas, the sort that is used for lorry sheets, made of cotton so that it swells slightly and becomes waterproof when it gets wet. I get mine from Green's Tarpaulines at Rawtenstall (01706 217019). Have it sewn up with the joins going from side to side, and the open edge of the join at the front – you will probably be stopped somewhere when it rains hard, with the back of the wagon to windward. Make sure that you measure the length needed to cover the wagon carefully, and get it slightly too large each way. Pull the edges down very hard when the canvas is dry, using upholstery pliers if you have got them, and fasten it down with a staple gun. Some people fasten the canvas down to a weighted board on each side, to stretch for a few days, before they make it permanent. I have usually been trying to get the job finished in time for Appleby, and haven't had that time to spare. Turn some of the spare canvas under the seats and inside the front and back bows and staple into place. Screw the weather boards on to the edges of the front and back bows, and onto the outside of the seats on top of the sheet. This being an 'Open Lot', there are a pair of canvas doors, made from the same material as the sheet. These should overlap slightly to keep out the wind and rain, and should hang right down below the floor level. I always have brass

Photo 11 With Sheet

eyelets put in all around the edges, and hook them onto brass cup hooks screwed into the front bow or front weatherboard. Pairs of eyelets, one on each door, let you lace them together for security. In general use, I use elastic rope with hooks at each end, the sort used for holding gear down on a roof rack, and hook the bottom of the doors to the shaft pin or something similar. When I am driving, I either take the doors right off and stow them inside, or fasten them back tightly so that they don't flap and frighten the horses. Sitting up on the front of a wagon with two horses bolting out of control is something you are not likely to want to experience twice!

Window. I usually just fill the window space with a flat sheet of plastic at first, held in place with half round strip tacked into place, making a good window later. A permanent window can project out so that you can see further around when looking out first thing in the morning to see if the tethered horses are alright. I like the window to have a small shelf over it on the inside, handy for putting a candle stick at night, perhaps a torch. If the window does project out like a miniature bow window, it is a good place to put a mug of tea. Lace curtains with a design of horses heads make a nice finish!

Cupboards. Most wagons have several cupboards built in, for storage of food, clothes and so on. I do without, storing things in removable boxes. I find that cupboards add weight and take up space, but you may very well decide to have them as a fixture. If you do, you will probably want to have a tall narrow cupoard right up at the front, level with the front board and front king posts on one or both sides. A flat top to the cupboard could well have a low rail around it to prevent things slipping off when on the move; any breakables should be wrapped up before you set off, anyway. Many wagons have a locker under the seats, with the door hinged to open downwards, and the space under the bed usualy has doors that open outwards, sometimes there are drawers that slide in under the bed above the seats. If you have room you may want to build a cupboard that is flush with the bed when that is pulled out, with the top at the right height to use as a table. I have thought for a long time of building a cupboard that is removeable, and when in use hangs on the side boarding, but somehow I have never got around to it. For the present I am happy with removeable boxes on the floor. However, I DO use stainless steel bread bins, the sort that have a door that opens at the front, with a flat top so that they can be stacked one on top of another, and I keep most of my fresh food in these, putting them on the front of the seats when I am stopped, and away under the bed when I move. I use a lot of brass cup hooks screwed into the runners to hang everything from baskets for fresh food, to spare ropes and harness. The pan box is often a second hand cupboard or chest, fastened under the floor behind the rear axle with long bolts or threaded rods. If you use a chest, it is more convenient to fit it so that it opens outward and downward.

Stove. They take up a lot of room, but do keep the wagon hot, not just warm, in winter. The classic type to have is the 'Queenie', or Queen Anne stove, with sliding doors at the front to refuel through, and a flat top with removeable rings for cooking, and on legs that can be fastened down to the floor, preferably on a sheet of metal to reduce the risk of the floor catching fire. There are several other good designs available, new and second hand, some of them really designed for boats, that are also short of space. To save floor space, the stove should be set back against the side, by cutting away part of the seat, and again protecting any woodwork that comes close to the stove with sheet metal. The chimney has to go out through the roof – make sure it and the stove are on the off or right hand side of the wagon to reduce the risk of hitting overhanging trees when on the road. You have to cut a hole in the roof for the chimney, and this has to be waterproof, or you will get a lot of rain coming in down the outside of the chimney. The best thing to do is have a sleeve made by a metal worker, welded onto a plate that is curved to fit your roof at that particular place, with a second plate of the same size and shape, or a little larger, to go inside the wagon, the one welded to the sleeve, outside the wagon, and each shaped so that the pipe from the stove fits tightly into the sleeve, and a short extension, probably with a rain proof cowl, fitting onto the top of the sleeve. When you install all of this, make sure that the lining, insulation and canvas sheet between the two plates are well away from the sleeve as it can get hot enough to set them alight. When every thing is in position, with a thin film of fireproof cement between the layers of the sandwich, the plates can be bolted tightly together. If the plates are just fastened to the canvas sheet, the chimney will tend to wobble about when on the move, so

Photo 12 Finished wagon on the road in 1997. Registered 13 year old Fell Pony gelding Hades Hill Oscar in the shafts with 7 year old grey mare Beth (out of a registered Fell mare, got by accident!) as 'side liner'.

make the inner plate wide enough to screw to a bow at each side, keeping the whole assembly steady. I have a small stove that burns charcoal efficiently in my full size wagon, but in the small one that I am now living in, I do without, using a small bottle gas stove or a portable charcoal or paraffin stove for cooking and heat in winter. In summer, especially when travelling, I use a fire on the ground outside for cooking. Most of the year round, the wagon gets so warm when the stove is lit that I have to leave the front sheet open. A stove takes up more than its fair share of space, what with having to keep things away from it, and storage for the wood, charcoal or coal fuel.

Now is the time to take off a lot of weight by **chamfering**, cutting away between joints and so on, reducing the cross section of the wood where it is safe to do this without losing strength, for example between the points where the runners and bows cross. and also along the weather boards, and it improves the look of the work. This can be done with a sharp chisel, or sharp knife, or with an electric rotary sander.

And all you have to do now is give all the woodwork a coating of clear preservative, and it is ready for painting. And living in.

Good luck up the road.

List of photos with foot notes

1. **Named parts**, viewed from front. Bows, runners, front crown board, rear crown board, window, king post, end deck, front former, side planking, bed frame, bearer, back board, seat, side, front board, choc rail, brake wheel, fore carriage, splinter bar, axle bed. This is an accommodation top that I built in 1993, on an eight foot by four foot Chesterfield dray built by Sonny Thorpe in 1986. Overall length at the seats eight foot eight, width at the seats, five foot four. Length inside from end deck to front former a bare six feet The bed extends from two feet nine to four feet five. Weight probably about seven hundredweight. I weighed an earlier, full size wagon, empty, and it was exactly ten hundredweight.

2. **Sides, seats and end deck** in position, rear crown not yet fitted. Note cratch or heck. The side board extends forwards of the floor

with a nice indented curve. This is an 'accommodation' type wagon, with the sides sitting on the floor, inside the choc rails and quarter rails.

3. **Looking at the end deck from inside**. Note poor fit with gaps on the seat! And the recess for the side planking to fit on top of the seat.

4. **Front and rear crown boards fitted**. The runners are stacked against the roof of the shed, ready for fitting into the slots cut in the crown boards. Note the planking of the end deck is recessed to take the side boarding. You can see the space for the porch between the front board and the front choc rail.

5. **The runners** are fastened down, but not yet cut to length.

6. **Front formers** fastened in position between seat and crown board

7. **Side boards** in place, still to be cut to front board to make it easier to step inside. Note recess in front former for side planking.

8. **Bows** all in place. Only 8 bows on this small (8 foot) dray. Bed bearers bolted on, bed frames in position. Runners and side planking cut flush with bows.

9. **View of rear** showing overhang. tilted slightly backwards.

10. **Inside** showing front crown board and king post, former, runners, bows and lining. Note how runners are recessed into crown board, a good fit. The head of the screw that fastens the top of the former to the crown board can just be seen. Some fancy curves in the centre of the crown board for ornament and more headroom, without losing strength.

11. **Canvas sheet** in position but not yet tightened and fastened down. Weatherboards still to be fitted. The porch can be clearly seen. The brackets under the seats have not yet been fitted. Note the empty milk crate used as a step! A nice curve on the laminated shafts.

12. **The finished wagon** on the road, painted, and with two ponies at a trot. At the time of writing, I have lived in this wagon for five years, and am about to set off in it to Appleby Fair for the sixth time.

Note on Restoration

This is much more difficult than building from scratch. Damaged parts have to be taken off very carefully and replaced acurately. Paint colours have to be matched, sometimes obsolete patterns of ironwork to be found. It will probably cost more to restore an old wagon than to build a good copy. Best not tried if you need telling how to do it.

Further reading

The English Gypsy Caravan, by C.H.Ward-Jackson and Dennis Harvey, David and Charles,1978

The Gypsies, Waggon- time and After, by Dennis Harvey, Barsford 1979 ISBN 0 7134 1548 7

Romani Culture and Gypsy Identity, edited by Thomas Acton and Gary Mundy, published University of Hertfordshire 1997, ISBN 0 900458 76 3, chapter 1. Gypsy Aesthetics, Identity and Creativity~ The Painted Wagon, by David Smith, Traveller Research Projects, Leicester.

Yorkshire Gypsy Fairs, Customs and Caravans by E Alan Jones, Hutton Press 1986 £4.75 ISBN 0 907033 43 1

Making Model Gypsy Caravans, by John Thompson, Fleet, Hampshire. Now out of print, this booklet, although intended for model makers, is so good that one can build a full size wagon from it.

Peter Ingram. Romany Museum at Selbourne, Hampshire, now closed. Articles in woodworking magazines on wagon restoration.

Woodworker December 1994, Bodger's eye view, by Stuart King, restoring a Showman's van

Wheelwrights for drays and wagons and repairs

Don Royston, Leeds 550135

Dave Peters, Preston, 01772 612220 Marsh villas Marsh lane, Longton

Frank Williams, Warton, Preston, 01253 762596

Philip Jowett, High Bentham, 01524 61626

Tony Clarke, Chinley, 50236

Edward Crowhurst, Southport, 212236

Joe Davidson, Blackpool,Messages to 01253 394065

Sonny Thorpe, Chesterfield, 01246 864861

Colin and Robert Dugdale, Melling, Lancaster 015242 21057

Painters

John Pocket, South Shropshire 01743 791568

John "Yorkie" Greenwood, Whitely Bay, Tyne and Wear, 0191 237 0855

Gill Barron, High Bentham, Lancaster 015242 62284

Katie and Sam. painting. Messages to Cheshire, 01925 241488, and Glocestershire,0122 253856.

Roy Morris, Bolton 391143

Horse Fairs

Lee Gap is one of the best places to find spare parts and bits and pieces, wheels, shafts, ironwork, items of harness, as well as horses and ponies, and sometimes even a complete wagon. First o'Lee is on the 24th August, and Latter Lee on September 17th. The Fair is at West Ardsley, a mile or two south of junction 28 on the M62. Appleby Fair is about a mile out of Appleby town in early June, Monday, Tuesday and finishing on the second Wednesday, 8th, 9th and 10th in 1998, followed by Bentham Fair on the Saturday.

Chorley Carriage Sales, twice a year, are another good source – here every thing is sold by auction from a catalogue – Smith Hodgkinson McGinty, Chorley, 01257 263633 (May 16th and September 26th in 1998).

Cutting List

Parana pine planking, 12 inch or wider, about half inch or five eigths after planing

Sides and Seats – one foot longer than dray, 4 off

Back and Front – width of dray, 2 off

Front and Back Crown Boards – ditto, 2 off

Front Formers – to fit between seat and crown, 2 off

Parana pine – one and three quarter to two inch square

King Posts – height of wagon, 4 off

Bed Frame – width of bed, 4 off

Bed Bearers – four foot six, 2 off

Parana pine – one and a half by three quarters

Runners – two feet longer than the dray, 5/7/9 off

Parana pine – one and half by a quarter plus

Bows – 14 feet or more, to fit, 10 or 12 off

Cedar tongue and groove planking – eg 4 inch wide, to cover three and half inch, half inch planed

End deck – to fill space eg over 6ft long, 12/14 off

Side planking – as long as runners eg 11ft, 16/20 off

Moulded edge planking – 4 inch by half inch, for weather boards, as long as seat eg 10ft, 2 off

Front and back weatherboards – cut to shape from eg 12inch by half inch planks, 4ft long, 6 off

Deal – 3" or 4" wide x 1"

Bed boards – 2'6" long

Prices

Quoted by James Ashworth Waterfoot (Successors) Ltd, 23/4/98

Parana 12" by 1" @ £2.16 per foot plus VAT

Parana 2" by 2" @ 75p per foot plus VAT

Parana 1½" by ¾" @ 30p per foot plus VAT

T&G redwood to cover 3½", ⅝ at 28p ft, ⅜ @ 1 p ft all plus VAT

Other materials:

Canvas, somewhere about 230 to 250 square feet for sheet, doors and cratch, 14 ounce

Screws, about 100 plus, No 8 by 1½" brass countersunk, 100 plus similar 1¼", all about £6 per hundred

Coach bolts, ¼", about two dozen

I use an electric drill, electric jig saw, cordless electric screw driver, and a GP hand saw, borrowing a mains socket and a bit of floor space when I need them. Going back a few years, many Travelling men built their own wagon out of whatever scrap wood and other materials they could find, using a saw from a tip, and a pocket knife, and then brought up a large and healthy family, while travelling and earning a living.

Cuprinol – I treat all parana pine with Cuprinol before painting.

PAINT YOUR

The decorations painted on Travellers' living waggons are based on the oldest doodle in the world – the Scroll – as used by Indians 5000 years ago, by Greeks, Romans, Irish, Aztecs, Africans ets., to decorate hand-made possessions they were proud of.

The scroll is a simple pattern that can fit into any shape. It can be plain or complicated, flowery or abstract, but to look right it should curve and flow as if it grew that way, and have a lively movement in it liek the plants it copies. With practice and confidence, a supple wrist and steady hand, it's a fast, flamboyant design. Like handwriting, every painter develops their own style. So although it's traditional, you can have fun with it and not be tied down to copying the "old masters". So have a bash, paint up your back door and the street a treat.

Brushes. Never mind A & C yet, B is what you want (see picture). "Sable writers" by Wright's of Lymm, from Trade or Art shops, will do thick or thin lines and strokes depending on angle you hold it and how you squash it. It's easier if you put your other fist under the one holding the brush, for steadiness, flexibility, and a clean lift-off. Sizes: depend on bird's quill used to bind hairs; lark, crow, duck or Extra Goose. Otherwise, a No.4 or 6 does most jobs, plus a fine for outlines. Cheaper versions by Daler or Proarte also work fine.

Paint. Whatever you've got in the right colours, which are: maroon, e.g. Woolworth, Romany, Post Office red, coach (dark green, middle (park-bench) green, gold, canary-yellow, and straw (cream). Please no mauves and day-glos. Keep's and Humbrol Enamels dry fast, no drips. Tekaloid is not good for this purpose. Undercoat is excellent. Varnish over the lot for an even surface. Ronseal Outdoor is OK on paint, doesn't crack up.

Thinners are most important, genuine turpentine works best. Work a drop into each brush-load to get the right flow. Keep in a "dipper", a bottle-top stuck down to pallette (glass, ply, biscuit tin lid) with bubblegum, and another for white spirit. Keep the colours clean.

All you need now is a cart or waggon to paint, but you'll probably have to build it yourself, so better get cracking.

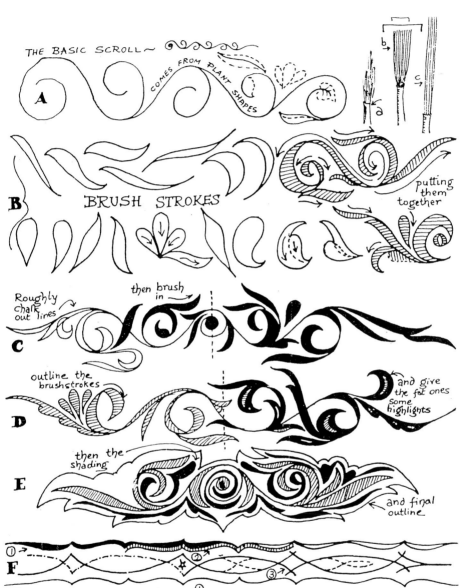

THE BASIC SCROLL~ COMES FROM PLANT SHAPES

A

B BRUSH STROKES

putting them together

Roughly chalk out lines — then brush in —

C

outline the brushstrokes

D and give the fat ones some highlights

then the shading

E and final outline

F ① ② ③

On carved work the chamfers ① (or scollops) are painted a lighter colour, or gilded with gold leaf, then edged in, ② and fancy line patterns added to fit shapes ③

© *Gill Barron 1982*

29